MEETING
the
ODDS

MEETING
the
ODDS

C.A. Evans

To order additional copies of this book, contact:
Xlibris Corporation
1-888-795-4274
www.Xlibris.com
Orders@Xlibris.com
96185

*To my dad and mom, thank you for
being a great example, your wonderful
personalities and can do spirit,*

*To my aunts and sisters,
thank you for your support, and
laughing with me,
I know this journey through the unkown
was hard on you too,*

*Kelby and Adrianne
you are the best gift ever!*

How vain it is to sit down
to write when you have not
stood up to live.
              Henry David Thoreau

NOTES TO THE READER

For the past four years I have tossed around the idea of writing this book. Back and forth with yes I need to tell this story and no I am not writing a book. From time to time I would tell people what had happened to me and that it happens around the world pretty much daily. Everyone would comment that I should write a book. Over time I realized that the story could reach more people in print than I could possibly talk to in a lifetime.

How do you write a book? Well I decided to put it on paper as though you and I were settled in for a long conversation. I wish I could be here to tell you the whole thing myself so you could get the facial expressions and laughter throughout. As painful as this year was, looking back and finding humor in these things is what helped me get through it all.

While I am a very ordinary person, doing ordinary things every single day, I have lived through some pretty extraordinary things. I had been in a mudslide in Colorado, the great '89 earthquake in California, and multiple floods and tornadoes in Oklahoma. So, if it is true that everything that happens today prepares you for tomorrow, well I was pretty well prepared.

You will find some statistics and references if you choose to read more on the subject. I never asked why me because if there are odds attached to something then there is a possibility you will find yourself meeting those odds one day. Thus the title of this book. The amazing thing about it is that I was not standing still but rather a moving target at 65 miles an hour! I had never thought about being struck by lightning, actually I believed that you died when that happened. I also thought you were safe in a vehicle. I suppose maybe you are if it is not running. There are a lot of details in this story, it is amazing that I could remember anything. So I hope that you find something useful in my story if not for yourself maybe you can help someone else.

CHAPTER 1

They say in Oklahoma if you don't like the weather wait a minute and it will change. The May 2007 Tornado outbreak began on May 4th and produced 84 tornadoes across Oklahoma, Colorado, Kansas and South Dakota. The severe weather continued for several days and finally left the area on May 7th, not to produce any more significant weather. Or did it?

Friday morning, May 4th, 2007, I went to work at 9 a.m. The law firm I was working for was particularly busy on fridays and I expected the day to go by fast. Storms were forming to the south and west all day so by noon I decided I should go home and pack my car so I would be ready to leave for Ponca City, Ok when my work day was done. I was looking forward to this weekend because I would be spending the entire weekend with my granddaughter who was turning three on May 8th. I missed Adrianne so much. I was offered the job at the law firm while I completed my bachelor degree and my aunt Vickie invited me to live with her until I graduated, so it was an offer I could not pass up. I went home at lunchtime and packed as much in my car as it would hold. Moving back to Ponca City was a temporary plan, until I graduated and found a job. After lunch I went back to work and the rest of the day flew by as it usually did. The office manager and the receptionist were fun ladies to work with and by five o'clock we were all ready for a weekend. As we walked out of the building to our cars we noticed the first thunderstorm was approaching from the southwest. We said our goodbyes and each went our separate ways.

As I was driving out of town I decided to stop and fill the car up with gas so I could drive straight through without stopping. I was about to finish filling up the gas tank, the wind was blowing hard now and the sky was very dark, when the tornado sirens began to sound off. I went inside to pay for the gas and the clerk said there was a tornado on the ground near Arnett which is west of Woodward. I got in my car and drove east out of Woodward knowing the storms were moving mostly north at this point. Generally, I like thunderstorms until they start dropping tornadoes. We have a lot of them in Oklahoma and tornadoes are to be taken seriously. By the time I drove the two and half hours home to Ponca City, OK there had been storms that dropped up to three inch hail and two tornadoes in the Ellis and Woodward County areas.

Saturday, May 5, 2007 the skies were clear in Ponca City and I took Adrianne shopping with me, I needed a new pair of walking shoes and she needed some too. She likes going shopping with Grandma, we always have fun on our trips to town. Adrianne was a special blessing for us, although she was born early and weighed only four pounds, she was healthy and strong. My mother is her Granny and I am her Grandma. My aunt Charlene stayed here with Granny from time to time and they all had great fun together. We picked up a birthday cake while we were in town and the family gathered to celebrate her birthday in the evening. From time to time through the day I checked the television to see the storms moving through the Woodward area. There had been rain with up to two inch hail and five tornadoes in that area by the end of the day.

Sunday, May 6, 2007 we continued to monitor the weather conditions out west, the storms had been rolling through all weekend and were not letting up. I was planning to leave about four o'clock so I could get back home and study. We spent the afternoon outside on the covered patio. Adrianne had grown since I had seen her last, she was now able to reach the top of the trunk of the car with her little fingertips. She kept going over to show me she could reach the antenna on my car. As time came near for me to leave I packed my car for the return trip home to Woodward. I had watched the weather patterns and decided maybe I could time it just right and drive into town in-between storms. Adrianne usually did not throw a fit when I left, but we had had a great weekend together and she was crying for me not to go. I always told her I would see her again soon and she usually was okay with that.

Well, I finally got on the road and stopped on the outskirts of town to get gas and a soda for the trip. I was just about to pull out onto the highway when my cell phone rang. It was my mother calling to tell me that my aunt had called from Woodward and said it would be better if I stayed in Ponca overnight because the storms were so bad and there was no break in-between where it would be safe to drive. She said that her son came by to check on her during the day and that they were all staying inside. I took her advice because we were both very independent women and generally did not let a thunderstorm stop us from going somewhere, so I was thinking it must have been really bad. I went back to my mother's house, unpacked the car and turned on the weather channel to see what was going on. There were thunderstorms moving through dropping more rain and hail up to one inch in diameter and it was expected to continue through the night.

It was early evening and Adrianne and my son had gone home, so I spread my books and papers out on the kitchen table. This last semester of school consisted of the capstone project, composing the documents to take a case the the Court of Appeals. An Appellate Brief as the final document. I had always wanted to complete my formal education and here I was already in the final project. Time had gone by so quickly and I had certainly enjoyed the journey. By nine o'clock I decided to pack it in for the night. I needed to get up early and a storm was to pass through our area in the night so I wanted to be rested for the early morning drive.

CHAPTER 2

    I arose early on the morning of May 7th planning to leave my mother's house by 5:30 am. Rising at 5:30am would allow me enough time gather my things on that morning and make the two and a half hour drive to Woodward, Oklahoma which is located in far northwestern Oklahoma. Once I arrived I would change my clothes for work which would start at 9 am and my work week would be off to a smooth start. Well, that was my plan anyway.

    After I arose I put on my favorite denim capri pants, a grey t-shirt and some flip flops. I made a pot of coffee, mom likes my coffee. I fixed my hair which was really easy to do thanks to my great hairdresser Shawnna, a little eye make up and some lip gloss and I was good to hit the road. As I was enjoying this morning, still feeling good about my weekend, I could hear thunder and rain outside, but it was moving through fast. I did not feel that I needed to rush around this morning; I was going to travel casual as I would have plenty of time to redress when I got to my aunt's house. At the bathroom mirror I started to put on my earrings and wrist watch and decided I would just go without them for the drive. As I walked back to the kitchen table I found myself starting to put them on again because I always wear earrings and a watch, but something was telling me not to. Feeling a little out of sorts, I tucked them in one of the many compartments of my "bling bling" purse. I loved that purse, it was just the right size. Chocolate brown, my color choice at the time. The purse had lots of compartments inside and out so I could find the items easily. And a lot of big crystals with a big silver buckle on the front of it. In hindsight it was so unusual for me to not wear those pieces of jewelry.

On this Monday morning, as I was gathering my things to put in my overnight bag, I thought I should go ahead and wear tennis shoes just in case I had car trouble and had to be out in the rain. I had purchased a new pair of tennis shoes while I was home for the weekend and I could break them in a little on the drive home. So I unboxed the tennis shoes, laced them up, put my life saving, feet soothing, orthodics in them and slipped on some of those great little no show socks and finally the tennis shoes. I was now sufficiently dressed and ready to load the car.

I packed everything in my bag and I walked straight out to the car and put everything in the trunk, my school books, coats and my overnight bag. It took a couple of trips but the car was loaded. I had brought my coats home this trip thinking I would leave them in Ponca City. However, I thought since we were having some pretty chilly days yet in May I would just carry them in the car a little while longer. When I got back in the house I remember thinking it was odd, I don't know why I put everything in the trunk because I usually threw everything in the backseat of my Ford Contour. As I poured another cup of coffee I was thinking it would surely help clear the webs out of my head. I sat down for a minute to talk to my mom and watch the weather report one more time to make sure there was no severe weather in the northern Oklahoma area that I would be driving into this morning.

We drank coffee while we watched the news and talked about the weather moving through our area and there should be no strong storms to navigate once I reached Interstate 35. And, by the time I turn back west to head for Enid it should be an easy drive the rest of the way to Woodward. With that said, I told Mom I was going to go ahead and take off. She jokingly said, "Don't let the lightning get you." Of course, I just giggled and went on out the door. Life was good and I was in good spirits most of the time. As I buckled in and started the car I remember thinking about which road I would take this morning. Since it was raining and a little stormy I decided I would take the interstate as it would be more traveled than Highway 177 South to the Cimarron Turnpike. Highway 177 South is a little two lane stretch, very dark and sparsley populated, that eventually takes you to Stillwater, OK best known for Oklahoma State University and Eskimo Joe's. About halfway between here and Stillwater you can access a turnpike that runs east and west. Going to the interstate would be a better choice considering the time of day and the weather. My dad, who shared the wisdom of his years and taught us how to take care of ourselves in this rough and tumble world, had always asked us to stay on well traveled roads when we were traveling in the dark or bad weather. Ninety-nine percent of the time his reasoning made sense.

Having let the car warm up and deciding what road to take, I was off but only drove about a mile up the highway and stopped in the Triple T convenience store for a cold fountain drink for the drive. With soda in hand I was now ready to go and the rain is steadily coming down but not bad at all to drive in. I headed west out of Ponca City on Highway 60 and got behind a semi truck about half way to I-35. It usually takes me about an hour and a half to drive to Enid from Mom's house but, it was going to take a little longer on this morning and that was okay, I had plenty of time.

I finally got to the interstate and of course the truck was turning on the on ramp too. Surely he would pick up speed when he got on the interstate. I slowed down a bit to get out of the trailing wind and rain coming off the truck and put my Chris LeDouix cd in so I would not have to keep trying to find a good station to listen to in this storm. I was making pretty good time at that point and it was still raining. An SUV passes me and gets in between me and the truck. Oh brother! The SUV slows down because now they are getting the wind and rain off the truck. I saw the rest stop up ahead and much to my surprise the truck and the SUV pulled off to the rest stop. Not much further and there is a truck stop and then just a little ways more and I will be turning off this southbound road and heading west to Enid.

I really thought I'd be out of the rain by the time I turn off for Enid. I finally took a few good sips of my diet coke and lit a cigarette. I passed under the overpass marking the Billings turn off and thought what a welcome site the truck stop was on the this dark, cold rainy morning. Just a little further and I will be out of the rain. I do love rain and thunderstorms but this had been a long weekend of violent weather and I was ready for some sunshine. I was also curious to see what kind of damage had been done in Woodward since they had been pounded with severe weather for days with rain, hail, winds and tornadoes. It had been one brutal storm cell after another.

Suddenly the rain in front of me caught my attention. It was so strange. The raindrops were very defined and they seemed to be lit up. At first I thought it was tiny bits of hail. But it wasn't. It was raindrops. That was the last thing I saw, the bright white little raindrops falling on the hood of the car and windshield.

CHAPTER 3

I opened my eyes because I heard laughter.

It was the sound of my laughter, soft and filled with joy. It felt as if I was coming back ever so slowly; floating into the present. I felt like I was wrapped in a warm blanket and did not want to wake up.

As I slowly opened my eyes, the inside of the car was dimly lit in shades of brown. Looking at the dashboard was odd; the inside of the car was usually lit in green, it was different now. I began to realize the inside of the car was filled with a brownish haze. I was not moving, just slowly moving my eyes around. My right hand was still on the steering wheel and my left arm was still on the arm rest of the driver's side door. I was still sitting upright and buckled in. Still holding the cigarette in my left hand that I had lit just a few minutes ago but, now it was burnt down to the butt. But how could that be . . . . I just lit it.

What in the world is going on? All the while this strange brown haze surrounded me I felt so peaceful and happy, not a worry in the world. My field of vision was pretty much straight ahead. I wasn't really aware (or concerned) about my total surroundings.

The airbags were out and deflated. I had never seen that before. There was a faint and constant sound that I can only describe as the sound of a jet engine roaring. Like when you are inside the plane preparing to take off. I finally looked up and out the windshield and realized the car was moving slowly along the shoulder of the road. I opened the door just enough that I could throw the cigarette butt down and when I looked down I could see that the car was to the right of the white line. I remember lifting my foot and gently touching the brake, the car stopped. I put my hand on the gear shift to put the car in park; the gear shift was in the console and very difficult to move now. I could not make the lever shift into park. I put the emergency brake on and took the key out of ignition. It was very dark and the light on the dashboard was getting dimmer as the brownish haze got thicker and thicker.

I fumbled around trying to find my cell phone that had been lying in the console next to me just a few minutes before and I could not find it. I reached over in the passenger seat, picked up my purse. The seatbelt was still fastened and so I unfastened the seat belt so I could lean over the console and find my misplaced cell phone. I wiggled around and was on my hands and knees feeling around on the passenger side of the car, I found the cell phone was on the passenger side floor as were a lot of quarters. I kept quarters in the console for the tolls and they were everywhere in the car now. I found this to be humorous and was kind of chuckling as I backed out of the car into the cool rain. It felt good, the rain on my face and to be in the cool air.

I stood there for a couple of minutes trying to get my bearings and I noticed the lights of the truck stop back down the road. I did not see a vehicle in sight either coming or going on the interstate and thought to myself this was odd for a busy interstate such as I-35. As I walked to the back of the car I noticed the backseat window was completely open. I opened the trunk. Smoke rolled out and was blowing all around me. I did not see flames, so I grabbed a couple of jackets and closed the trunk lid. I put the heaviest jacket on and threw a lighter one over my head because I was going to have to walk back to the truck stop I'd passed by. It was then an odor hit me that was like burning electrical wires, melting plastic and the odor you would get when someone is welding.

I stood there and dialed 911. My phone would not dial a number. The screen was lit up so I tried to dial the Highway Patrol; maybe one of them would be nearby. My cell phone would not dial. I know I was pressing the numbers, the screen on the phone would not show the numbers but it was lit up. Then I noticed headlights coming from the north so I started waving my phone to maybe flag down the truck. A semi truck blazed past me, did not even slow down.

Still standing in the rain behind the car, the headlights and tail lights were still on, smoke is drifting all around it and I realized a semi truck just went past me at seventy miles an hour and I did not hear it. I looked to the side of the road and noticed mile marker 203. I am at mile marker 203. I cannot hear the numbers on my phone being dialed and no numbers are showing up when I press them. Maybe I am dead. Maybe I am not really standing here. I can feel the rain on my face but I can't feel anything else.

I don't feel my purse hanging from my arm, I don't feel the phone in my hand. Thinking was so very hard to do. It seemed as though my mind and body were not working together! I put my cell phone in my pocket. What is happening?

CHAPTER 4

There was no point in standing by the car in the rain any longer so I looked to see if there were any vehicles coming from the north, of course not. I crossed the first two lanes and then there was the grassy, wet and muddy median. I remember thinking I hope there are no snakes in here, and I slowly made my way down the slope, across the mud and up the other side. I looked to make sure there were no headlights coming from the south, and there weren't any so I crossed on over the next two lanes and to the shoulder. It seemed like a very long ways across the pavement and as I stood there on the shoulder of the interstate I looked back at the car. It must have been really hot because there was still smoke surrounding it but oddly the headlights and tail lights were still on. The smoke coming from the car was not like anything I had seen before when I had had car troubles. It was then the western sky lit up with red lightning and it occurred to me that perhaps lightning had hit my car and more so, me.

As I looked back to the east I saw some Black Angus cows in the pasture next to the road. That was comforting to me as they had been a part of my life since I could remember growing up in Oklahoma. The sky beyond the pasture lit up with red lighting and so I started walking north toward the truck stop.

It is funny what went through my mind on this walk. I felt as though my body was over here and another part of me was over there. Strange feeling. And everything seemed to be moving in slow motion. I would walk a little and then a thought would come to me, things like, "Is this it, is this my final walk on earth?"; "Well here I am walking in the rain, red lighting all around."; "If I get struck by lightning again on this walk surely it will kill me instantly."; "Here I am walking in the rain with a big ole bling bling purse and oh, I am carrying my keys in my hand but I can't feel them."; "I can't hear anything so maybe I am dead."; "When I get to the truck stop and talk to the clerk if they don't answer then I will know for sure I am not really here." I kept walking and walking and thinking to myself or, I may have been talking out loud and just could not hear it, "I can't even feel my feet hitting the pavement and it is a lot further to that truck stop than it looks!" Even though it was a real effort to keep my mind and body together, as one, I was determined to make it to that truck stop and see if anyone could see or hear me!

As I approached the exit ramp that would eventually get me to the truck stop, I looked back one more time to see my car. I could barely see the tail lights. The car was a deep crimson color and kind of hard to see in the dark anyways. About halfway into the exit ramp a car passed by rather slowly, but kept on going. I just kept walking, thankful for everyone and everything in my life. Especially for the great weekend with my family, getting to be with Adrianne my precious granddaughter a little while longer yesterday, and for driving me to stay in shape this year because it was sure was paying off now! I had been taking walks in the evenings to lower the stress of my last semester of school and all the unknowns ahead of me trying to find a job and relocate again. The walking trails in Woodward, Oklahoma are amazing. You can go up long steep grades, winding paths through their agricultural research station, you can go off the trail and walk through the cemetery (alot of people walk around there), or just enjoy walking along the roadside. This small rural town has great paved walking trails.

Finally, I made it up and around the exit ramp and still had a ways to go to get inside the truck stop but I just kept walking. As I was walking across the parking lot I took the jacket off my head and looked around to see the people getting gas. Some of them appeared to be noticing someone walking across the lot! They were probably wondering who would wear capri pants, a heavy coat and a jacket thrown over their head, carry a big shinny purse and walk on the interstate, in the rain at this time of day. Oh my gosh! Believe me, I had never imagined anything like this.

CHAPTER 5

I made it inside the building and there were a couple of people at the counter paying for their fuel. I took my cell phone out of my pocket and tried to dial my mom's phone number. The phone still would not dial. I tried again and again while I was standing there waiting to get to the counter. It could not have been more than five minutes but I really was not aware of much still. A young man with a safety vest on walked up to me and asked if I was alright. I told him my car had been hit by lightning up the road, my cell phone would not work, I had walked up here and I would really like to make a couple of phone calls but I can bearly hear. I remember he was a tall, thin young man with brown hair and he was so kind. He just raised his hand as if to say, okay or wait here.

Now the people inside the truck stop were standing around looking at me, a half dozen or so by now. I don't know if I was talking loud, or maybe they just overheard the conversation and were curious. I walked over to the side of the counter and the young man in the safety vest quickly returned with some paper towels and put one over my right hand as he lifted it up and laid it on the counter. There was some blood on my right hand but I could not look at it yet. I was feeling weak, but determined. I don't think I had even blinked since I opened my eyes in the car. He stood there holding my arm while I gave the lady at the counter my mother's phone number. Maybe they thought I would pass out or something, I was just thankful to be inside with people. She handed me the phone. I told Mom my car had been hit by lightning and I was calling a wrecker and I would be back home later. I said "I'm fine, don't worry". And then I hung up because I could not hear anyway and I did not want to talk because then she would know I was not alright.

Next, I asked the lady behind the counter to call the Highway Patrol. I cannot stress enough how kind these two people were in the truck stop, she told the person that answered the call what happened and that I was in the truck stop. Now, you can laugh with me on what I was directed to do next because I found it to be hilarious. The lady at the counter was relaying the questions and answers and I was pretty much reading lips, hearing was not happening.

Operator: Are you alright, do you want an ambulance?
Me: No ambulance.
Operator: Are you sure.
Me: Yes, I am sure.
Operator: Do you need a wrecker?
Me: Yes, call Warners in Ponca City.
Operator: You need to go back and stay with your car.
Me: What!?
Clerk: They want you to go back and stay with your car.
Me: I just got hit by lightning in that car, I walked clear up here in the rain and lightning, Warners will pick me up here, I am not going back to the car!

The operator wanted to talk to me so I took the phone. She wanted to call an ambulance for me but I told her nothing was broken, my hand was bleeding a little and I would wait here for the wrecker.

For some reason the general rule must be to stay with your disabled car. I don't know the reasoning but, there are exceptions to the rule, and I was not going to walk back down there in this condition, in the rain and sit in a smoldering car by myself, waiting for a wrecker.

The lady at the counter asked if I would like to have something cold to drink and use the ladies room. I did need something to drink so the young man got me a diet coke and walked with me to the restroom. I went inside and took a couple of big gulps before walking over to the sink where there was a big mirror. I don't remember tasting the coke, just that it was cool and refreshing. I had not even looked at my right hand to see where the blood was coming from. Obviously it was not a gaping wound and there was not a great blood loss going on. And I could not feel anything. So I put my drink down on the counter and hung my purse and the jacket I had over my head on the stall door. Keeping my head down not to look in the mirror. I pulled the paper towel on my right hand off slowly so I could slowly see what was the matter. My right thumb looked like a huge chicken leg. There was blood coming from a small hole where the knuckle bone was. The whole side of my hand from the index finger down around the thumb and clear to the palm of my hand was shades of yellow, green and black. Strange though, I could move all my fingers and thumb, it did not hurt at all. Okay, well then. I stood in front of the mirror and finally looked straight ahead. My face, and neck looked like I had a nice sunburn. There is a spot over my left eye and a streak across my forehead that is a deeper shade of red. I ran my left hand through my hair, all over my head. No blood or pain up there. I turned the water on and washed my hands and face with soap and water. I got some paper towels wet and wiped down my legs. They did not have the sunburnt look. Wiping my legs off just seemed like the thing to do. Feeling a bit relieved, I combed my hair with my left hand and made it look as good as it was going to right now. It was strange not feeling anything, not even my arms moving. My coordination was a little off too, like if you are right handed and try to do something simple like brush your teeth with your left hand. I gathered everything up and a wet towel for my hand and went back out to the seating area.

The seating area inside the building was along the west side of the building, lined with large windows. My legs felt weak and I was determined to hold it together so I put the jacket that was over my head on the back of the chair and sat down. As I sat there looking out at the rain thinking it should be getting light soon I looked in the direction of my car. I could not see my car down the road; the landscape is a little hilly in this part of the country so the car was now out of my view. It occurred to me that I needed to call my aunt, she would be expecting to see me early this morning. I dialed my cell phone again. This time it dialed, but just as quickly it disconnected. Call ended. Okay, I dialed it again, this time it rang through. Wow, I said hello. Hello? I know I heard her saying hello back a couple of times in my left ear. I don't know exactly how the conversation went except that I told her my car was struck by lightning, I had a few bumps and bruises but could she please call into work for me this morning as I would not be back in Woodward for a few days. I told her I was at a truck stop waiting for a wrecker and would call her back later because this was the only call that had gone through on my cell phone. I hung up because my voice was a little shaky and I really did not want anyone to worry. There is nothing worse than being a long ways away from someone that could use your help. I was so relieved to have gotten a call through on that phone. I think a tear may have rolled down my cheek at this point.

There were a couple of men sitting in the same area that were staring in my direction. I remember looking at each of them for a couple of minutes wondering if they had been inside the store since I arrived and what they were thinking. The young man in the safety jacket was now outside the side door near where I was sitting. I took my purse and drink and went outside. The rain was but a mist now and it was light outside. I walked a little ways away from where he was mopping the entrance outside and lit a cigarette. He smiled and I must have said something funny because we both had a good laugh. I was getting restless, it seemed like it had been a long time since we called for the wrecker and I was hoping I would not have to wait here for hours.

I went back inside and asked the lady behind the counter to please call the wrecker to see how long I may be waiting. She promptly called the wrecker service in Ponca City and reported that they had dispatched a truck and it should be there shortly. The man on the phone said something to the tune of be patient, the driver was called out early and was on his way. I went back to the table I had been sitting at and sat in a chair where I could watch for the wrecker coming.

I don't know how long I sat there but is seemed like a very long time and I finally saw the huge red wrecker pulled in and parked in front of the entrance to the truck stop. I met the driver at the door and told him I was ready to go if he was. He was a young man with blondish hair and medium build. I climbed into the passenger seat and as we fastened our seat belts I proceeded to tell him that I could not hear so I would have to look at him if he needed to talk to me. I asked him to head south on the interstate. We were off, it was raining again. We drove a mile or so and he looked at me and asked where is it? It was kind of funny. "Just keep going we will find it", I said.

Soon we were at the car and the headlights and tail lights were still on. That was amazing to me, the car would not run, the ignition had been turned off for hours but the lights were on. He pulled up to the front of the car and backed the truck up. He jumped out and went to work. I stayed inside and tried to watch through a very large side mirror. I really could not see anything because the car was right behind the truck, as it should be. The driver got back in the driver's seat and was a little out of breath. I was watching him warming his hands and he said he was finally able to get the gear shift where he needed it. He looked at me kind of funny, like he was confused, I really did not know what to think. He then began to operate some controls in the truck, I was still so very peaceful, just there. He got out again and when he returned this time he said we were ready to go. I turned around and looked over my left shoulder, out the rear window of the truck and remember seeing the headlights of the car up above the truck. The only wrecker I have seen around town is the kind that picks up the front of a vehicle tows it down the road. But with this one the car is up on top of the bed of the truck. Well isn't that something.

The driver looked out his side mirror and started pulling the wrecker toward the median. I remember saying something to him about it being really muddy in there and maybe we should not take the wrecker in there. (The way my day seemed to be going, getting the wrecker stuck in the bottom of the muddy median was a great possibility.) He was very confident and said we would be fine. I may have closed my eyes at this point because I remember feeling the truck sliding a little and then we were up on the other side of the interstate and as I looked over at the driver he looked pretty proud of his driving skills.

At last, we were heading north on Interstate 35 and would be back in Ponca City in about thirty or forty-five minutes. When we turned off the interestate to Highway 60 East I dialed my phone. Still no luck. The ride home was quiet and I was starting to feel pain on the left side of my head and my right hand had a burning sensation. It was making me a little nervous starting to feel pain and I just wanted to get home, get cleaned up and rest. All I could smell was something like burning electrical wires melting plastic and hot metal and thought I just needed a shower and clean clothes. I also thought I should call my insurance agent and report my car so I could get that process started. The driver pulled up to the office door and let me out. He was going to another building where my car would be kept.

As I went inside I looked at their clock, it was now going on 9 a.m. I sat down at the desk and gave the gentleman my information. I apologized for not being able to hear but we got through the process. He sent me down the hallway to another room where another man asked me some questions. I remember working so hard to hear or understand what they were asking for. It was funny, this man could not hear well either. He stood in front of me and appeared to be talking really, really loud while pointing to his ear telling me he could not hear, I needed to speak up! After I signed papers and they got their copies made I went back to the front desk and sat down. I remember asking the gentleman to use the phone to call for a ride. He looked straight at me and said, "you have a cell phone in your hand." I looked straight back and said, "yes, but it does not work, see." He finally smiled. I proceeded to tell him that I was a personal friend of his late brother and our children had grown up together and I needed to go just about a mile from their house. He sat there at the desk looking down for what seemed to be an eternity and finally he motioned to one of the men standing in the office and told him to take me where ever I needed to go. I remember feeling really tired and just wanting to get home.

I don't know how it came about but I was taken to the building where my car would be kept. The wrecker truck was backed up to the door with my car still on top and several men were standing around it talking. I remember one of them telling me how lucky I was this morning, I wasn't feeling it at the time. The next thing I remember we were pulling up the driveway to my mother's front door.

CHAPTER 6

Mom had the front door open and was waiting for me in the living room. Of course the first question was are you alright and then what happened? I know we had a conversation and she was feeling really overwhelmed and said she needed to lay down. I was feeling really tired and hot and realized I still had my heavy coat on. I took my coat off and hung it over a kitchen chair, it smelled just like my smoldering car so I did not want to put it in the closet. I had not taken my coat off since I put it on in the rain this morning. This is when I realized that I had left my other jacket at the truck stop. It was a really pretty tailored tapestry jacket and I really did like it, I was sorry to have left it behind. Mom had walked into the bathroom and I was going to go tell her that I was going to take a shower and lay down myself. On the way there I noticed the burn on my right arm. I stopped and looked at it for a minute. It was very bright red, about the width of my index finger. It wrapped around my right arm from the wrist down around the inside of my arm to the elbow. It was really strange. it was not a blistery looking burn like I have had when I burned my hand on the stove, and there was bruising on both sides of it. I remember just standing there looking at it, it was so hard to think, nothing about this day was making any sense to me. It seemed like I had been through so much this morning and the day had been moving in slow motion.

So I walked on into the bathroom, Mom was standing at the sink looking in the mirror. I held my arm up and calmly told her I thought I would go on in to the hospital and get checked over. I was was feeling a little woosy and bless her heart, she never does well in stressful situations and this one was making her sick. I told her to lay down and relax, don't worry. She offered to drive me into town but I knew that I was not getting in a car with her driving at this time. Looking back, I had no business driving either, but I did.

I made it to the hospital emergency room and sat in the waiting room for awhile, watching the rain and the lady mopping the floor. According to the documents I arrived at 10:05 am and was in triage at 10:07 am. The next thing I remember was laying on the bed, one person taking my vitals, another was asking me questions. It was so hard to think. The top of my head hurt. My left jaw hurt. I could not hear very well so they had to be facing me to talk. My eyes were burning. When I took my t-shirt off to put on the hospital gown we found another big ugly burn all the way across my chest. I still felt numb and felt no emotions at this point and that was a good thing.

As they went about their business a tech came in with her tote to obviously take blood. When she walked in the curtain I noticed she had pretty long red hair, she said, "Seth?". I looked at her and said, "who". She said, "Seth?". I said, "No, I am not Seth." She then called my name and I said, "yes, I am Charlotte". She said they needed to take blood samples and as she proceeded I laid there wondering why she would call out the name Seth. My former neighbor must have heard about the accident this morning at work because he worked at the paint and body shop next door to the wrecker service which is owned by the same people. He may have called my mother's house after I left and found out I was at the hospital. It was just odd that the tech would come in my room and call out his uncommon name.

Next, a nurse told me they were going to run an EKG. So they all left and I laid there with the curtain drawn, for what seemed like a long time when two of them came back with the machine. I had not been a patient in a hospital emergency room for many years, but having previously worked in this particular hospital for seven years made this whole process a little easier to take. They ran their test and periodically someone would come in a take my blood pressure and ask how I was doing. It was hard for me to lay on my back any longer so I sat up on the edge of the bed. They had me wash my hands with soap again to see if the burning sensation would let up at all. They brought me a coke. Everyone I came in contact with kept telling me how lucky I was and things like alot of people drop dead hours after a lightning strike, even six months afterwards. That was information I could have done without. I was cleared to put my clothes back on and a nurse took my blood pressure again before I could leave. It finally occurred to me I should ask what my blood pressure was and she said something like 200 over something. It was hard to think, even I knew I was not going anywhere with that kind of reading and I told her so. She said it had been up and down since I came in and she would come back in a few minutes and take it again. So I sat there on the edge of the bed, wanting to go home if they could not do anything more here. The nurse returned and took my blood pressure again, it was 164/95 and explained that if I felt anything funny in my chest, like my heart racing I should come back in. She then went over the doctors orders that included: watch for dark urine; watch for palpitations of the heart; watch for severe muscle aches, watch for numbness; and drink lot of fluids. I was given the name and phone number of a doctor to follow up with. I was released at 11:40 a.m. When I got back in the car I sat there and read the orders again and had to laugh just to release some tension because I was already numb all over, how could I feel heart palpitations in my chest anyway and I had the kind of blood pressure readings most people go to the hospital for but I was being released to go home and rest with no meds. How funny is that.

CHAPTER 7

When I got home Mom was sleeping and I went in and laid down. The day was creeping by, I could not fall asleep so I decided to go get my belongings out of my car. It was about four o'clock in the afternoon and the wrecker office would be closing soon. The rain had finally stopped and the sky was clear so I found a box in the garage and drove to the wrecker company. I went in the office to let them know I was there and they had a man go with me to the building where my car was. I backed the car I was driving up to the big door of the building while the man went inside and opened the big door and turned on a light. He pointed me in the direction of my car, there were a lot of cars in there.

As I walked up to my car I noticed the windshield had two huge starburst breaks in it. I was talking out loud the whole time, the man that escorted me was a very good listener who stood there with his hands in his pockets and would throw in a comment now and then. I walked around the front of the car to the passenger side in amazement, looking at the windshield. How could I have not seen these before now? Then I walked around to the driver side and opened the door. The keys were in the car and one of the things I wanted to retrieve was my key ring. I sat down in the driver's seat to take the key off the key ring and was looking around. The more I saw, the more I realized that I was a very lucky lady today. Sitting in the driver's seat I notice the gear shift in the console is laying over to the side. I reached over and picked it up, it had been blown out of the console and was hanging by wires. Then I looked around and started picking up quarters, they were in the front seat, on the floor and in the backseat and the floor. Wow. As I was picking up quarters I also picked up the plastic air vents. How odd is that, they were blown out of the dash. There were burn marks on the carpet. The backseat looked charred. I found my umbrella laying between the front passenger seat and the door, it had been in the back seat this morning when I left. After I boxed up everything from inside the car I carried that box out to the car and breathed in some fresh air. I went back in to retrieve my books and overnight bag from the trunk.

Still so very amazed that I had not seen any of the damages to my car, and as I walked to the back of the car I noticed the only window that was up was the drivers window. I put the key in the keyhole to open the trunk. Now I noticed that there was a big hole to my right, in the back of my car. I think you could put a soda can in that hole! I was still having trouble comprehending this whole situation. I stood there looking at the hole thinking that was where my radio antenna was. Then I looked a little further up the back of the car and saw that my radio antenna was gone, but the base was still there. A little burnt, but the base was there. Feeling very confused, I looked back at the hole and the paint all around the hole is burned away at varying degrees. A rather large area. I said to the man with me, "how could I have not seen this before now?". I was a little apprehensive about opening the trunk, would my books and papers be burnt, what was I about to see now? I wanted to get this done and be on my way. I opened the trunk and the lid blocked the light from above so I could not see anything inside, just shadows. I asked for a flashlight and the gentlemen said he did not have one. I got my box with my school books in it, an overnight bag, my jack, a first aid kit and a fishing pole. I carried them one by one to the car outside and put them in that trunk. The man that was there with me watched me coming and going. He laughed when he saw the fishing pole come out of the trunk, but hey I like to fish and my son had given me this one. With everything loaded in the other car I thanked the gentleman and headed back home.

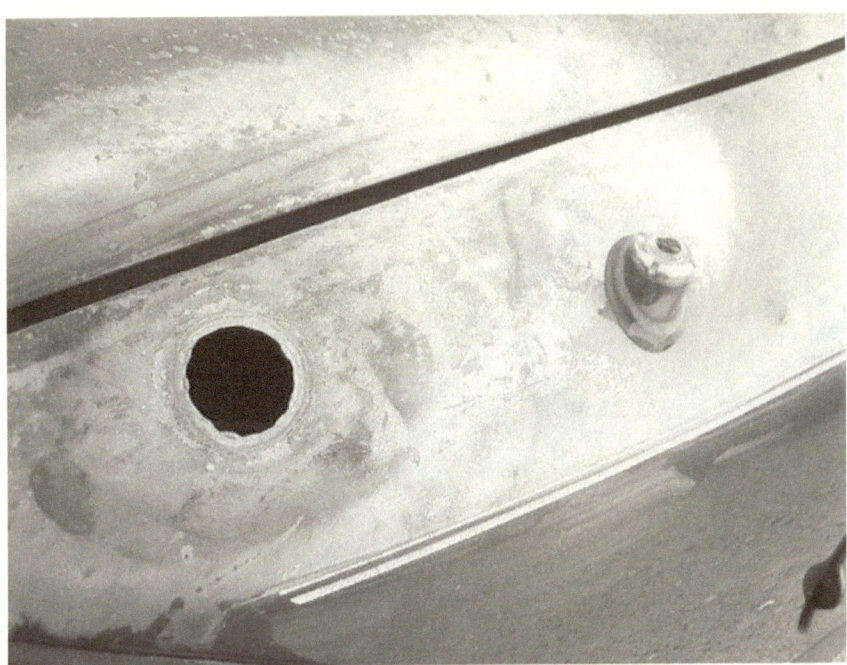

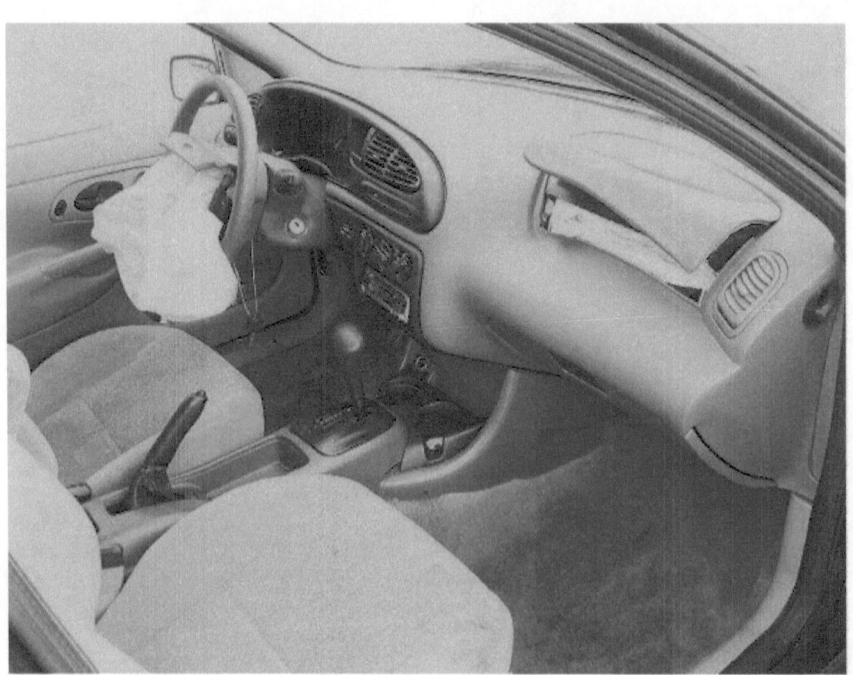

I have no recollection of when I spoke with my insurance agent this day, but I did. I was finally going to take a shower and get out of these clothes. I sat down to take my shoes off and was looking at them to see if they had been damaged at all. They looked fine, other than being dirty from walking around in mud and water all day. When I took my shoes off I found there was blood on the sock of my left foot, the top of the toe of the sock about the size of a quarter. I just sat there looking at it, not really wanting to take my sock off. I felt no pain in my foot, I wiggled my toes and then I just sat there on the edge of the bed for a minute. I took my right sock off first and looked my foot and toes over real good, everything was fine there. So then I took my left sock off, the bleeding had stopped but there was a tiny hole on the top of the knuckle of one toe. I wasn't a scrape or just missing skin, it was a hole.

I have no other recollections of anything the rest of the day other than finally getting to take a shower, treating all the wounds and borrowed a night gown from my mother. I don't remember eating anything all day. I could not tell you when it was that I went to bed but that is where I woke up the next morning.

The following day, May 8, 2007, I recall talking with my insurance agent Tina, whom I had known for many years. She told me that she had been an agent for twenty years and had never had anyone that had been struck by lightning. She said the adjuster had looked at my car and it was definitely a total loss. It had not occurred to me to lift the hood of the car and take a look in there. It was probably best. The insurance company later sent me photos and I can see why they considered the car a total loss.

The only other memory I have of this day was calling the doctor I was supposed to follow up with today. First the phone number I was given was not a working number. Not too comforting, but I guess it happens. Secondly, when I did reach the office the receptionist asked what I needed to be seen for and how I would pay for this visit. I replied that I had been struck by lightning the day before and the emergency room had given me this doctor for a follow-up and that I had coverage through my auto insurance for the medical payments. She was sorry that had happened to me but the doctor could not see me as he does not take this type of patient! Okay, well then, I will see my doctor in Woodward, OK. I would liked to have seen a doctor and maybe under other circumstances I would have called around and found one, but I wasn't thinking clearly obviously and I just wanted to rest. I thought I would be strong enough to go back to Woodward by the end of the week.

I was able to sleep only a couple of hours at a time and then I would be wide awake. My head felt very hot on the top and across my forehead. My son came by and we talked about what had happened and what I was going to do. Other than that, I don't know what I did in-between periods of sleep.

Wednesday, May 9, 2007, I was very stiff and sore. Bruises had come out on my ankles, the muscle on the side of the right foot and up the fronts of my legs; my head hurt, it felt full of pressure, hot and heavy and; my hair, my hair had patches that stood on end! Root Booster doesn't keep my hair standing up like that all day. The hair on the top and upper sides of my head actually stood up. I had eye pain, actually my eyeballs just plain hurt, and my left jaw and right ear really hurt. The streak across my chest was now red, purple and green. The marks on my forehead remained red and I still looked like I had a nice sunburn. My right hand was now black and green. The skin on my forearms looked as if it had been twisted. I rationalized that it too would heal in time. If I had seen a nurse today and was asked to rate my pain, it would have been a nine or ten. I seldom take any kind of medication but I was ready for something extra strength.

This was now day three, I remember talking with one of the attorneys at the law firm and telling her I would drive back to Woodward on Thursday and come to work on Friday. She was fine with that. I recall talking with my aunt and my mom and son but that is the extent of the memories of Wednesday. Still determined, nothing to do but keep moving and in time everything would heal. I needed to get back to life.

CHAPTER 8

I got around about midmorning Thursday, May 10, 2007, slowly. Mom was going to let me use her car for the remainder of the month, as I would be moving back here with her soon. I still had pain and was lacking energy but I knew I had to make the drive to Woodward. It was around noon when I left, I wanted to stop by and see my doctor in Woodward. I had my papers from the hospital emergency room with me.

I was doing fine driving, the weather was great and I remember having the window down enjoying the fresh air. As I approached the Billings exit I started feeling funny. As I passed under the overpass I looked over at the truck stop that I had walked to just a few days earlier. From that point my body began to sweat, I mean wringing wet, sweating bullets. I was not panicking, I did not feel emotional. Apparently somewhere in my brain there is a memory and my body apparently was reacting to this place. I was concentrating on the road and driving carefully as I passed by mile marker 203. My body was so hot, sweat was running down the sides of my face, my clothes felt wet. After driving a little further and taking the exit to Enid I felt relieved that I had made it this far and everything settled down. I had made it through that rough spot and I believed eventually I would be able to pass by here without any problem.

When I got to Enid I stopped at the first gas station and got out to walk around and cool off. I figured if I could use the restroom and get something cold to drink it would be nice break before driving the second half of the trip. Everything went smooth the rest of the drive and I was back in Woodward in time to go by the doctor's office. Wouldn't you know it, the doctor was out the remainder of the week, but they scheduled me to see him on Monday.

I drove to my aunt's house, it was now early evening. She and her daughter and grandkids were sitting outside enjoying the weather. I walked up and chatted just a couple of minutes as they kind of looked me over. It was if they did not really want to look at me, and that is okay I didn't really want to either. I went inside to my room, turned on the tv and laid across the bed. I was not sleepy, just tired.

I went in to work on Friday morning trying to get back into a normal routine under the circumstances. Everyone at the office was glad it was friday, they had endured storms for days and were tired too. We traded stories throughout the day, I was glad I had not been here through the storms and they were glad they were not in my shoes too! The one thing that I hoped I would not have to do today was transcribe a taped meeting. And, as it turned out that is the one thing I did most of the day. Because my hearing was impaired and it was my first full day of being active, it took a lot of foot pedal work to get through the tapes but I made it!

Over the weekend I ate, slept and worked on my school project. I was very frustrated that I was having so much trouble thinking. Usually I could read and grasp the material, but now I was having to concentrate on every single thing I read. Writing was equally as difficult. I remember being frustrated to the point of sitting at the table with my head in my hands just praying to get better and being able to complete this project. This project was what I needed to graduate and I did not want to take this semester over! The middle of May was here and I needed to have everything on paper so I could edit the content and format down to every comma and period. I thought I would just take a break from the project and read a book. I had made it a habit to read a biography every month, so I took the latest purchase out on the patio with some tea and read for a couple of hours. Reading other people's stories about how they came to where they are now is uplifting. Everyone has troubles and successes, but it seems that the bigger the successes are so are the troubles. Anyway, taking the breaks from the school project turned out to be a good thing, I could go back for shorter intervals and get more done in the end.

Monday, with papers in hand I told the doctor what had happened and he looked me over and he too told me how lucky I was. He was concerned with the black coloration of the bruises and that he did not see a particular test on my papers so he called the Ponca City Hospital to see if it had been done and it had not. The doctor had the nurse take the blood and they took it over to the lab immediately as it had been seven days since I had been struck by lightning. He said to go home and he would call me later tonight. I had done everything I needed to today and now I was just waiting for the doctor to call which he did. He called later in the evening, and said the test came back good. He reminded me to take it easy, drink a lot of fluids and come back if I was not feeling well or did not feel I was improving. So, that was a relief.

As the submission date neared for my school project I was hanging in there at work and slowly moving my things back to Ponca City on the weekends. My hearing had either improved somewhat or it had became my new normal.

The pain in my jaws was a great concern and I wanted to get it checked out. I contacted the oral surgeon in Ponca City that I had used over the years and made an appointment that resulted in a broken jaw tooth being removed. I felt better in a couple of days and was glad I had removed the tooth. Usually a dentist does not want to remove teeth, just as a last resort. I had worn braces for three years, and had my wisdom teeth removed as a teenager. I did not even have a cavity or another tooth removed until I was in my forties!

Now June and the check came in for the loss of my car. At the same time another big storm came through Woodward dropping hail that nearly totalled my mother's red Grand Prix that I was using! I hated to call and tell her the news.

With my final project completed and submitted and a date set as the last day at work it was time to make the final trip back to Ponca. I was actually feeling pretty good other than getting tired easily.

CHAPTER 9

It was now the first of July, I am back home and taking a little break to get settled in before seriously hunting for a job. I had come to the conclusion that with no car and all of my money needed for medical bills that there was no way I could make the trip to Chicago for the graduation ceremonies. It was not the end of the world, I was very proud that I had finished my formal education and the ceremony was just icing on the cake. I had my diploma mailed to the house and moved on.

I had all the dents taken out of my mothers' car from the hail storm, it looked good as new.

Another tooth broke and I was in the dentist office getting the remainder extracted so I might enjoy the coming July 4th weekend.

I spent the summer gaining strength with bouts of gallbladder pain and kidney stones. By September I took a part-time job at a casino, counting money. I was having difficulty standing for nine to twelve hours a day so after a couple of months I took a job at the local Wal-mart in the cosmetics department. I was like a kid in a candy store! I do love working in retail and helping the customers. I continued to feel stronger but the bruises on my ankles and chest remained, the sides of my face would turn very dark green and black at times, and I continued to lose a tooth now and then.

I needed something to do other than work. I had been a trail guide and volunteered in the gift shop at the Tallgrass Prairie for five years before I decided to complete my education. Working fulltime and school fulltime left little me time so I decided I would volunteer again when I had completed my education and was settled in somewhere. I had inquired two years earlier about becoming a volunteer with the local CASA program. Being a CASA I could continue to serve just about anywhere I lived. Now that I had graduated and was working it was time to sign up. I went to the local office to get the paperwork started. The director and I had worked at the same local hospital for a number of years and it was good to see her. We had more in common than I could have imagined. As it turns out she is a paralegal too. We had a really nice talk and in the course of conversation she said she had a short term job that would end in the spring if I was interested. We worked out a schedule and everything seemed to be falling into place. The extra money would help me with my doctor bills as I had no insurance, and I would learn more about the CASA program before becoming a volunteer.

I was purposely keeping busy . . . it works for me. Adrianne came to stay with us nearly every weekend. The time spent with her makes all your concerns take a backseat. She would be awake to tell me goodbye in the mornings and ready to play in the evening. When I had a tooth pulled and was laid out in the recliner she would be the nurse. In November 2007 I took a full time job with Walmart. The holiday season was very busy. The CASA office was equally busy with holiday activities and preparing for a new year.

CHAPTER 10

Before I knew it spring was here and I was signing up to take the classes necessary to become a CASA. I was able to arrange my work schedule so I could attend the classes that would end in April.

On April 10th, 2008 I was at the CASA office working. We had some pretty good thunderstorms this week, but this day, all day, the wind blew hard and I felt restless. I was planning on going home for an hour or so before my evening class. Mid afternoon I had the feeling that I needed to go home. I tried to keep working and I called home to see if everything was okay. Mom said they were fine, but the feeling would not go away. I could hear the wind howling outside the brick building and suddenly I found myself standing in front of my desk telling the people in the office I needed to get home and I would be back later for my class. I picked up my purse and drove home.

The wind was terrible. I parked the car under the carport so no blowing little tree limbs would damage the car. I left my purse in the car knowing I would not be home long. Mom, my aunt and Adrianne were all in living room and I sat down and talked about the crazy wind. I told them I just felt like I needed to come home so I did and I would go back later for my class. They thought I was silly, they were fine and there was no need to come home early. Mom decided to go out and get the newspaper. Her paper box is at the end of the driveway. I laughed as she went out the door because she is not too steady on her feet anyway and in this wind I could not believe she wanted to walk out to get the paper.

It seemed she had been gone too long and there was a lot of debris hitting the house from the trees. I still had the jitters for reasons unknown, so I walked out the front door to the end of the carport to look out and see if she was making it back down the driveway. There she was, standing out in the driveway looking up at the big cottonwood trees blowing and swirling. I yelled as loud as I could in the terrible howling wind for her to come on in the house. She walked as fast as she could against the wind and when I saw she was almost to the carport I went back in and sat down. I was really stressed out at this point and sat down facing the front door. She walked in slowly, brushing her hair back off her face, saying, "she was just looking at the trees and did not know why I was worried."

As she laid the newspaper on the table where I was sitting I looked out the front door. I did not know what was happening but it looked like a tree was coming right toward the house! It made me dizzy in that split second seeing that huge tree twirling and I could not say anything, I jumped up and grabbed Adrianne who was playing in the middle of the floor. Boy was she mad at me. I think I scared her because I picked her up and ran for the back of the house. We made it about a hop, skip and a jump before the tree hit the house. It hit hard and made all kinds of noise. We were all taken by surprise. It was a huge old tree. The grannies did not see it coming so they were really confused.

A couple of neighbors came running over yelling mom's name with eyes as big as quarters. They had seen her go out to the paper box but did not see her make it in the house. She and I went to the door and saw that the her red Grand Prix was smashed under the carport with the huge tree resting on top of it all. The front storm door would open but it was not safe to walk out under the carport so she talked to the neighbors through the door assuring them everyone was alright.

The winds passed and settled down within an hour. Mom was on the phone with her insurance agent reporting the damage. Reality had set in and she was trying to keep her composure. She was nearly killed, within seconds. And, had the car not been parked in front of the house under the carport the tree surely would have came crashing through the house where all three of them had been sitting this afternoon.

We went outside and walked around looking at what had just happened. Oddly, mom had the tree beside this one removed just a week or two earlier because it had dropped limbs and damaged the carport several times. She had also had the one that just fell examined to see if it needed to be removed and this particular tree was deemed healthy and not a threat!

The tree laid all the way across the yard, and over the house. Mom's car was crushed under the carport. It was definitely totalled now. My purse was in the car. A neighbor gave me a camera to take pictures with and the newspaper reporter came out and took pictures as well. People were driving down our road all evening looking at what had happened. I called the office and told them what had happened and that I would not be in this evening for class. Adrianne's mom came and picked her up later in the evening. I looked in the attic and sure enough there were tree limbs poking through the roof on the front side of the house.

We listened to the tree and the carport moving all night. It was frightening. It was a couple of days before all the insurance adjusters made it out. We were exhausted from the sounds of the broken carport and the tree. There had been heavy storms in other parts of the state this week and they were very busy. When the adjuster came to look at the car he got my purse out for me. With the car being totalled we were able to get a rental car for three days and mom hired a company to remove the tree from the house and take the carport away as well. The roof and the front of the house would be replaced. Most importantly, we were all safe.

I was back at work after a couple of days and as luck would have it when I returned to work there was a car posted on the bulletin board for sale. Dependable and cheap. I took mom to look at it and she bought it.

By the end of the month I had completed my CASA classes. I took the Oath of Office on April 28, 2008. It was a good day and I would get a case soon.

It is now May 2008, Adrianne celebrated another birthday and it is Mother's Day. I had worked all day and was sitting at the table late in the evening enjoying some chocolate my son had given me. And can you believe it, two teeth broke in half, so I sat there spitting out chocolate and my teeth! Oddly the teeth that broke were on opposite sides of my mouth. So back to the dentist I went and had the remainder of those teeth removed.

It had been a year since I was struck by lightning driving down the interstate. Many things had happened. I had been in the doctor's offices more this past year than I had been in my entire life. The year 2007 started out good and I was looking at many new beginnings for this past year. Looking back it was more like a series of survivor!

> You only live once—
> but if you work it right,
> once is enough.
> Joe E. Lewis

CONCLUSION

At the time of this writing the four year anniversary of being struck by lightning is approaching. I still like the sound of rain and thunder, but do not like driving in it. I do it only when necessary and if there is a lot of lightning I will call in to work and be a little late or if I am leaving work and it is storming I will hang around until it lets up.

As far as my health, it is good overall, here is an update:

I still have a numbness on the top of my head and there are areas where the hair just will not grow out longer than it was on the day I was struck.

In the right ear I hear my heartbeat or it may be the blood passing through the veins, regardless it is heard 24/7.

My eyes are very sensitive to temperature change, they water a lot.

The scar is still across my chest as well as other marks on my right arm and a spot over my left eye.

I still have no feeling in my chest. I thought it would return over time. I had always had a healthy heartbeat and I could feel the blood rushing through my chest, but now I feel nothing.

My ankles are still bruised, have red spots and are painful on a daily basis.

I ended up losing seven teeth, and four more are loose. I have no jaw teeth on the right side and a couple left on the left side. I thought the loose ones would tighten up over time but they haven't. Dentures are in my future.

I do not get the feeling of being full when eating so I have to watch the intake.

I am able to eat a wide range of foods, just nothing too rich or greasy though. My gallbladder malfunctioned a couple months after being struck by lightning and let me tell you it is one organ you want to keep happy.

I also had two episodes with kidney stones in the first year. They make you cry, childbirth was not that bad. If I drink a lot of water I have no problems there, dehydration is a battle.

Sleep is better as time goes on. I can sleep even though I roll around and wake up wrapped in the sheet most nights. I still have periods where I cannot sleep all night long. After missing a whole night of sleep I still function well the next day and am able to sleep the second night.

My right arm and hand are sometimes shaky. A friend refers to it as tremors. I don't know what brings that on, it never lasts long and thank goodness for that.

It is not all gloom and doom. I believe that I should also tell you that I have not had headaches. The traditional sinus pounding headaches that I had lived with all my life. I have not had any headaches or sinus pain since being struck by lightning.

I don't feel pain in my shoulder, or back. I used to have lower back pain, and pain deep in the back side of my right shoulder. The kind of pain I experienced in my back and shoulder sent me to the chiropractor and a local doctor at times for electric shock therapy.

All things considered I feel very blessed and I hope being free of headaches, back aches and shoulder pain is a permanent condition.

> Think of the beauty that is
> still left in and around you
> and by happy!
> Anne Frank

THE END

WANT MORE INFORMATION

I can give you a little information about lightning that I found as I was trying to figure out what in the world hit me. There's just no way to figure it out really, it was just lightning. But, lightning comes without warning. It is so unpredictable, we never know where or what it will strike. Historically, lightning kills more people than any other natural event only outranked by floods. The National Weather Service has changed it's slogan to "When thunder roars, go indoors". Don't stand outside counting to 30 to see where the bolt is, go indoors and count to 30!

Keraunomedicine is the medical study of lightning casualties.

Keraunopathy is the study of the effects of a lightning strike rather than treatment.

Lightning travels as fast as 100,000 miles a second. Channels have been observed longer than ten miles in length. Lightning strikes with ten million to 100 million volts with currents up to 50,000 plus amps. The experts on the subject say this is four to five times hotter than the sun. This happens in a split second also referred to as a millisecond.

Hot lightning (high-current lightning) lasts more than a second. The high energy results in melting and/or carbonizing large objects.

The heat of a lightning bolt can burn tissue, cause lung damage, and the chest can be damaged as a result of the forces of expanding hot air.

Electricity generally causes cardiac arrest. It may also flash over the body leaving burns or Lichtengerg figures that can last for hour or days.

Ruptured eardrums are most common and ocular cataracts can develop more than a year afterwards.

Blunt trauma injuries occur from the shock waves when the surrounding air expands and implodes.

For more information regarding lightning and safety go to the websites of the National Weather Service and Lightning Strike Survivors. Another good read is a paper titled "Lightning Injuries" by Eric L. Johnson, MD.

Odds of Becoming a Lightning Strike Victim

U.S. 2000 Census Popluation 280,000,000

Odds of being struck by lightning in a
> given year (reported deaths + injuries) 1/700,000

Odds of being struck by lightning in your
> lifetime (Est. 80 years) 1/3000

Odds that you will be affected by someone
> being struck (10 people per victim) 1/300

Just because you have been struck once does not take you out of the pool of possibility. There are people that have been struck multiple times so far in their lifetime. And since every person who survives a lightning strike has a set of unique injuries and after effects, treatment is difficult to find. Don't give up. Get in the support group. Talk to other survivors. There are some things medicine can help. There is help for lightning strike survivors and their families needing medical and legal assistance at Lightning Strike Survivors.com.

Don't be discouraged.
It is often the last key
in the bunch that opens
the lock.

Anonymous

ACKNOWLEDGEMENTS

This book could not be complete without thanking each and every person who assisted me on May 7, 2007 and the days that followed:

The employees at the Conoco Truck Stop, Billings Exit I35;
Warners Wrecker Service, Ponca City, OK;
The Oklahoma Highway Patrol;
Emergency Room Staff, Ponca City Medical Center;
Tina Brown, Loftis Wetzel Insurance Agency;
Dr. Cao, Woodward, OK;
Ray Kinsinger (deceased), Oral Surgeon Staff;
Dr. J. Russell Hill and Staff, Urgent Care & Family Practice Center;
Victor Andrews, DDS, MAGD, and Staff Family Denistry;
Shawnna Burgess, Hair Designer;
And of course, my family.

Thank you to my sister Kristi who helped me get started writing, insisted on more details and supported me through the process. Thank you to my mom, who kept me fed and reminded me I needed to sleep when I worked too late, and my son who took care of the yard and anything else I needed while I locked myself in with my computer to get the manuscript done.

Then I need to say thank you to all the people at Xlibris Publishing for their support through this project. Thank you Sergio Lee, Publisist for getting the project on the drawing board. Thank you Kenneth Niemes, Submission Advisor great knowledge base. Thank you Joy Asis, Author Services for calling me constantly to see where I was and when I would be finished! Joy worked with me on my work schedule and writing schedule and kept me moving and meeting those deadlines. And thank you to all the behind the scenes professionals at Xlibris that designed the cover and interior layout and the marketing team for getting the book out to the readers. What a great team.

Best Wishes,

Charlotte A. Evans

SERENITY PRAYER

God grant me the
Serenity to accept the things I cannot change;
Courage to change the things I can; and
Wisdom to know the difference.
Living one day at a time;
Enjoying one moment at a time;
Accepting hardship as the pathway to peace;
Taking as He did, this sinful world as it is,
not as I would have it;
Trusting that He will make all things right
if I surrender to His Will;
That I may be reasonably happy in this life
and supremely happy with Him
Forever in the next.

Amen.

Rienhold Niebuhr